NUREG-1611

Aging Management of Nuclear Power Plant Containments for License Renewal

I0502968

Manuscript Completed: September 1997
Date Published: September 1997

W. C. Liu, P. T. Kuo, S. S. Lee

Division of Reactor Program Management
Office of Nuclear Reactor Regulation
U.S. Nuclear Regulatory Commission
Washington, DC 20555-0001

ABSTRACT

The Nuclear Regulatory Commission (NRC) published its license renewal rule, Title 10 of the Code of Federal Regulations (10 CFR) Part 54, on May 8, 1995, providing the requirements for renewal of operating licenses for nuclear power plants. 10 CFR 54.21(a)(1)(i) requires an aging management review of containment structures to ensure that the effects of aging will be managed so that their intended functions will be maintained for the period of extended operation. In 1990, the Nuclear Management and Resources Council (NUMARC), now the Nuclear Energy Institute (NEI), submitted for NRC review, the industry reports (IRs), NUMARC Report 90-01 and NUMARC Report 90-10, addressing aging management issues associated with PWR containments and BWR containments for license renewal, respectively.

Recently, the Commission amended 10 CFR 50.55a to promulgate requirements for inservice inspection of containment structures. The final rule on §50.55a, "Codes and Standards for Nuclear Power Plants; Subsection IWE and Subsection IWL," was published in August 1996. This rule incorporates by reference the 1992 Edition with the 1992 Addenda of Subsections IWE and IWL of Section XI, Division 1, of the American Society of Mechanical Engineers (ASME) Boiler and Pressure Vessel Code addressing the inservice inspection of metal containments/liners and concrete containments, respectively.

The purpose of this report is to reconcile the technical information and agreements resulting from the NUMARC IR reviews and the inservice inspection requirements of Subsections IWE and IWL as promulgated in §50.55a for license renewal consideration. This report concludes that Subsections IWE and IWL of Section XI, Division 1, of the ASME Code as endorsed in §50.55a are generally consistent with the technical information and agreements reached during the IR reviews. Specific exceptions are identified and additional evaluations and augmented inspection activities for renewal are recommended.

CONTENTS

TABLES

APPENDICES

ABBREVIATIONS

ACI   American Concrete Institute
AISC   American Institute of Steel Construction
ARDM   Age-Related Degradation Mechanism
ASME   American Society of Mechanical Engineers
ASTM   American Society for Testing and Materials

BWR   Boiling Water Reactor

CS   Carbon Steel
CFR   Code of Federal Regulations
CRD   Control Rod Drive

ECCS   Emergency Core Cooling System

GSI   Generic Safety Issues

IR   Industry Report
ISI   Inservice Inspection

IWE   Subsection of ASME Code, Section XI, "Rules for Inservice Inspection of Nuclear Power Plant Components," containing "Requirements for Class MC and Metallic Liners of Class CC Components of Light-Water Cooled Plants"

IWF   Subsection of ASME Code, Section XI, "Rules for Inservice Inspection of Nuclear Power Plant Components," containing "Requirements for Class 1, 2, 3, and MC Component Supports of Light-Water Cooled Plants"

IWL   Subsection of ASME Code, Section XI, "Rules for Inservice Inspection of Nuclear Power Plant Components," containing "Requirements for Class CC Concrete Components of Light-Water Cooled Plants"

NEI   Nuclear Energy Institute
NRC   Nuclear Regulatory Commission
NUMARC   Nuclear Management and Resources Council

ppm   Parts per million
PWR   Pressurized Water Reactor

RG   Regulatory Guide

SS   Stainless Steel
SCC   Stress Corrosion Cracking

LIST OF TABLES

NUREG-1611

TABLE 1. AGING MANAGEMENT OF PWR CONTAINMENTS FOR LICENSE RENEWAL

Table 2. AGING MANAGEMENT OF BWR CONTAINMENTS FOR LICENSE RENEWAL

1.0 INTRODUCTION

Part 54 of 10 CFR, the license renewal rule, was published on May 8, 1995, providing requirements for renewal of operating licenses for nuclear power plants. 10 CFR 54.21(a)(1)(i) requires an aging management review of structures and components within the scope of license renewal to ensure that the effects of aging will be managed so that their intended functions will be maintained for the period of extended operation. Containment structures are subject to this requirement.

Recently, the Commission amended 10 CFR 50.55a to promulgate requirements for inservice inspection of containment structures. The final rule on §50.55a, "Codes and Standards for Nuclear Power Plants; Subsection IWE and Subsection IWL," was published on August 8, 1996 (61 FR 41303). This rule incorporates by reference the 1992 Edition with the 1992 Addenda of Subsections IWE and IWL of Section XI, Division 1, of the American Society of Mechanical Engineers (ASME) Boiler and Pressure Vessel Code addressing the inservice inspection of metal containments/liners and concrete containments, respectively [References 1 and 2]. Guidance for implementation of the containment inspection requirements is described in Appendix A of this report.

In 1990, the Nuclear Management and Resources Council (NUMARC), now the Nuclear Energy Institute (NEI), submitted for NRC review, ten industry reports (IRs) addressing aging issues associated with specific structures and components of nuclear power plants for license renewal. Of the 10 IRs, one addresses PWR containments, and another addresses BWR containments [References 3 and 4]. No safety evaluations were developed for the review of these IRs. However, NUREG-1557 provides a brief summary of the technical information and NUMARC/NRC agreements resulting from the review of nine of the 10 IRs. The NUMARC Report 90-08, "Low-Voltage, In-Containment, Environmentally-qualified Cable License Renewal Industry Report,"[Reference 5] is not addressed in this NUREG since the subject is being addressed under GSI-168.

On August 26, 1996, the Commission issued Draft Regulatory Guide DG-1047, "Standard Format and Content for Applications to Renew Nuclear Power Plant Operating Licenses." for public comment as part of the implementation of 10 CFR Part 54, the license renewal rule. A comment was received concerning whether the NRC staff had any plans to limit the scope of license renewal review for containments since the final rule on 10 CFR 50.55a, which endorses the 1992 Edition with the 1992 Addenda of Subsections IWE and IWL, was published on August 8, 1996. The purpose of this NUREG is to reconcile the technical information and agreements resulting from the NUMARC IR reviews and the inservice inspection requirements of Subsections IWE and IWL as promulgated in §50.55a for license renewal consideration.

2.0 LICENSE RENEWAL EVALUATION OF AGING MANAGEMENT OF CONTAINMENT STRUCTURES

The NRC staff reviewed Tables B3 and B4 of NUREG-1557 [Reference 5] for the PWR and BWR containments, respectively, to determine if Subsections IWE and IWL inspection requirements are consistent with the technical agreements from the IR reviews. Where NUREG-1557 indicates that an aging effect on specific structures is non-significant, the NRC staff recommends no aging management program. Where NUREG-1557 indicates that an aging effect is non-significant if certain conditions are met, the NRC staff reviewed Subsections IWE/IWL and §50.55a to determine if the containment inspection requirements would be adequate to manage that aging effect regardless of whether those conditions are met. Where NUREG-1557 indicates that an aging effect should be managed with specified programs, the NRC staff reviewed Subsections IWE/IWL and §50.55a requirements to determine if they are adequate to manage that aging effect for the renewal term. If the NRC staff determined that Subsections IWE/IWL and §50.55a requirements should be augmented to manage a certain aging effect for the renewal term, additional inspections or evaluations are recommended. The results of the NRC staff evaluation are provided in Tables 1 and 2 of this report for the PWR and BWR containments, respectively. The PWR and BWR containment structural components evaluated in NUREG-1557 are listed in Appendix B and Appendix C of this report, respectively.

3.0 CONCLUSIONS

The NRC staff has reconciled the technical information and agreements from the NUMARC IR reviews and the inspection requirements of Subsections IWE/IWL as promulgated in §50.55a for managing the effects of aging for PWR and BWR containments for the period of extended operation. The staff found that the requirements of Subsections IWE/IWL and §50.55a will be an effective aging management program for managing the aging effects of containment structures for the period of extended operation, provided that the following additional evaluations and inspections specifically for license renewal are also performed:

a. Specific requirements contained in Part 54, the license renewal rule, such as Part 54.4 for scoping and intended function, and Part 54.21 for evaluating time-limited aging analyses.

b. ASME Section XI, Appendix VII and Appendix VIII [References 6 and 7] to be implemented when ultrasonic examinations are utilized for inspection of containments.

c. The following issues, in addition to implementing Subsections IWE/IWL through §50.55a, should be addressed in a license renewal application:

   (1) Management of potential aging effects of structures in inaccessible areas when conditions in accessible areas may not indicate the presence of or result in degradation to such inaccessible areas. This is discussed in Items 1, 3, 4, 9, and 13 of Table 1 for PWR containments, and in Items 1, 3, 4, 9, and 13 of Table 2 for BWR containments of this report.

   (2) Fatigue associated with containment penetration bellows and penetration sleeves. This is discussed in Item 16 of Table 1 for PWR containments, and in Item 16 of Table 2 for BWR containments of this report.

(3) Settlement associated with containment concrete basemat bearing on soil or piles, or experiencing significant changes in ground water conditions. This is discussed in Item 17 of Table 1 for PWR containments, and in Item 17 of Table 2 for BWR containments of this report.

(4) Erosion of cement for porous concrete if subfoundation layers of porous concrete are used in the construction of containment concrete basemat with the presence of underground water. This is discussed in Item 18 of Table 1 for PWR containments, and in Item 18 of Table 2 for BWR containments of this report.

(5) Performance of examinations specified in Examination Category E-B for pressure retaining welds, and Examination Category E-F for pressure retaining dissimilar metal welds of Subsection IWE for license renewal. This is discussed in Item 12 of Table 1 for PWR containments, and Item 12 of Table 2 for BWR containments of this report.

(6) Cracking of penetration bellows. This is discussed in Item 12 of Table 1 for PWR containments, and Item 12 of Table 2 for BWR containments of this report.

(7) Elevated temperature of prestressing tendons for (prestressed) concrete containments. This is discussed in Item 14 of Table 1 for PWR containments, and Item 14 of Table 2 for BWR containments of this report.

The NRC staff recommends that the requirements of Subsections IWE and IWL through §50.55a, and those items identified in sections 3.a. through 3.c. above be incorporated into the Standard Review Plan for License Renewal (SRP-LR).

REFERENCES

1. Subsection IWE, "Requirements for Class MC and Metallic Liners of Class CC components of Light-Water Cooled Power Plants," Section XI, Division 1, Boiler and Pressure Vessel Code, American Society of Mechanical Engineers, New York, N.Y., 1992 Edition and 1992 Addenda.

2. Subsection IWL, "Requirements for Class CC Concrete Components of Light-Water Cooled Power Plants," Section XI, Division 1, Boiler and Pressure Vessel Code, American Society of Mechanical Engineers, New York, N.Y., 1992 Edition and 1992 Addenda.

3. "Pressurized Water Reactor Containment Structures, License Renewal Industry Report," NUMARC Report Number 90-01, Revision 1, Nuclear Management and Resource Council, September 1991.

4. "BWR Containments, License Renewal Industry Report," NUMARC Report Number 90-10, Revision 1, Nuclear Management and Resource Council, December 1991.

5. NUREG-1557, "Summary of Technical Information and Agreements from Nuclear Management and Resources Council Industry Reports Addressing License Renewal," U.S. Nuclear Regulatory Commission, October 1996.

6. Appendix VII, "Qualification of Nondestructive Examination Personnel for Ultrasonic Examination," Section XI, Division 1, Boiler and Pressure Vessel Code, American Society of Mechanical Engineers, New York, N.Y., 1989 Edition.

7. Appendix VIII, "Performance Demonstration for Ultrasonic Examination Systems," Section XI, Division 1, Boiler and Pressure Vessel Code, American Society of Mechanical Engineers, New York, N.Y., 1989 Addenda.

8. NRC Information Notice 97-10, "Liner Plate Corrosion in Concrete Containments," March 13, 1997.

9. NRC Information Notice 97-11, "Cement Erosion from Containment Subfoundations at Nuclear Power Plants," March 21, 1997.

10. NRC Regulatory Guide 1.35, Revision 3, "Inservice Inspection of Ungrouted Tendons in Prestressed Concrete Containments," July 1990.

11. ACI 201.2R-77, "Guide to Durable Concrete," American Concrete Institute.**

12. ACI 215R-74, "Consideration for Design of Concrete Structures Subjected to Fatigue Loading," American Concrete Institute.**

13. ACI 318, "Building Code Requirements for Reinforced Concrete," American Concrete Institute.**

REFERENCES (Continued)

14. ACI 359, "Code for Concrete Reactor Vessel and Containments," American Concrete Institute.**

15. NRC Information Notice 92-20, "Inadequate Local Leak Rate Testing," March 3, 1992.

16. Letter from T. F. Plunkett of the Florida Power and Light Company to Stewart D. Ebneter of NRC, dated January 25, 1993.

**: denotes that the citation is used as a reference to provide only background information.

TABLE 1. AGING MANAGEMENT OF PWR CONTAINMENTS FOR LICENSE RENEWAL

Item	Component, Aging Mechanism & Aging Effects	Issue and Evaluation[*]
01	Concrete & Steel Containment Aging mechanism: Not applicable. Aging effects: General	Issue: A "one-time inspection for license renewal." NUREG-1557 states that one time inspection is an unresolved issue regarding staff request for inspection of concrete containment & steel containment to assess the current condition of containment and to provide a baseline information for any future inspections (Page B-28 of NUREG-1557). Recommendation: The issue is resolved with the implementation of IWE/IWL through §50.55a. However, specific-recommendations for applicable aging effects are addressed hereinafter in this table. Discussion: Subsections IWE and IWL require periodic inspection of the containment in accessible areas. These inspections would periodically assess the condition of the containment and each inspection would provide a documented baseline for subsequent inspections. Furthermore, §50.55a(b)(2)(ix)(E) and (b)(2)(x)(A) require an evaluation of inaccessible areas when conditions exist in accessible areas that could indicate the presence of or result in degradation to such inaccessible areas. However, the management of potential aging effects of inaccessible areas, when conditions in accessible areas may not indicate the presence of or result in degradation to such inaccessible areas, is addressed individually for each applicable aging effect (i.e., Items 1, 3, 4, 9, and 13 of this table). Conditions for such inaccessible areas should be evaluated for license renewal. A program for a one-time inspection may be proposed.

| 02 | Concrete Structure

Aging mechanism:

Freeze-thaw

Aging effects:

Scaling, cracking, & spalling | Issue: NUREG-1557 states that freeze-thaw is non-significant if the following conditions are met: concrete containment structures located in a geographic regions of negligible weathering conditions (weathering index <100 day-inch/yr); and if located in severe (weathering index >500 day-inch/yr) or moderate (100-500 day-inch/yr) weathering conditions with the concrete mix design meets the air content & water-to-cement ratio requirements of ASTM C260 or equivalently, the ASME Sect. III, Division 2, paragraph CC 2231.7.1. The issue of whether freeze-thaw is potentially significant for the concrete containment dome, particularly in severe weathering regions, is identified as unresolved (Page B-29 of NUREG-1557).

Recommendation: The issue is resolved with the implementation of IWL.

Discussion: Freeze-thaw results in scaling, cracking, and spalling. Any freeze-thaw degradation would initially appear in the exposed concrete structure. Subsection IWL, Examination Category L-A, requires periodic visual examination of accessible concrete surfaces and would detect any freeze-thaw damage of the concrete containment, including the dome, regardless of whether the above weathering conditions are met. |

| 03 | Concrete Structure

Aging mechanism:

Leaching of calcium hydroxide

Aging effects:

Increase of porosity & permeability | Issue: NUREG-1557 states that leaching of calcium hydroxide is non-significant for containment concrete structures if the following conditions are met: concrete structures not exposed to flowing water; and for concrete structures that are exposed to flowing water but are constructed using the guidance of ACI 201.2R-77 to ensure dense, well-cured concrete with low permeability and control cracking through proper arrangement and distribution of reinforcement (Page B-29 of NUREG-1557).

Recommendation: The issue would be managed with the implementation of IWL through §50.55a. However, the management of potential leaching of calcium hydroxide of inaccessible areas of containment concrete structures when conditions in accessible areas may not indicate the presence of or result in degradation to such inaccessible areas needs to be justified on a plant-specific basis.

Discussion: IWL, Examination Category L-A, requires periodic examination of accessible concrete surfaces and §50.55a(b)(2)(ix)(E) requires an evaluation of inaccessible areas when conditions exist in accessible areas that could indicate the presence of or result in degradation to such inaccessible areas. Regardless of whether the above conditions are met, potential leaching of calcium hydroxide would be detected as water stains on accessible surfaces by the IWL visual examination. However, the management of potential leaching of calcium hydroxide of inaccessible areas (e.g., below grade portion of concrete structures with presence of flowing water) when conditions in accessible areas may not indicate the presence of or result in degradation to such inaccessible areas, needs to be evaluated on a plant specific basis. |

| 04 | Concrete Structure

Aging mechanism:

Aggressive chemical attack

Aging effects:

Increase of porosity and permeability, cracking, and spalling | Issue: NUREG-1557 states that aggressive chemical attack is non-significant for above grade concrete containment structures because they are not exposed to ground water. Aggressive chemical attack is non-significant for below grade concrete containment structures if the following conditions are met: containment concrete is not exposed to aggressive ground water (pH <5.5, chloride >500 ppm, & sulfate >1500 ppm); or if exposed to ground water that exceeds the pH, chloride, sulfate limits, the exposure is for intermittent periods only. NUREG-1557 indicates that inspection of concrete containment structure should be in accordance with IWL. NUREG-1557 states that evaluation for management of inaccessible areas of below grade concrete containment structures is to be justified on a plant-specific basis (Page B-30 of NUREG-1557).

Recommendation: The issue would be managed with the implementation of IWL through §50.55a. However, the management of potential aggressive chemical attack of inaccessible areas of containment concrete structures when conditions in accessible areas may not indicate the presence of or result in degradation to such inaccessible areas needs to be justified on a plant-specific basis.

Discussion: Aggressive chemical attack results in increase of porosity and permeability, cracking and spalling. IWL, Examination Category L-A, requires periodic examination of accessible concrete surfaces and §50.55a(b)(2)(ix)(E) requires an evaluation of inaccessible areas when conditions exist in accessible areas that could indicate the presence of or result in degradation to such inaccessible areas. Regardless of whether the above conditions are met, potential aggressive chemical attack would be detected by IWL and §50.55a(b)(2)(ix)(E). However, the management of potential aggressive chemical attack of inaccessible areas when conditions in accessible areas may not indicate the presence of or result in degradation to such inaccessible areas needs to be evaluated. |

05	Concrete Structure	Issue: NUREG-1557 states that reaction with aggregates is an unresolved issue. NUREG-1557 indicates that the NRC staff believes that alkaline-aggregate reactions can not be ruled out. Tests involving aggregates alone are not satisfactory in predicting aggregate performance. Alkaline-aggregate reaction may occur after 25 or more years (Page B-31 of NUREG-1557).
	Aging mechanism:	
	Reaction with aggregates	Recommendation: The issue is resolved with the implementation of IWL through §50.55a.
	Aging effects:	Discussion: If alkaline-aggregate reaction occurs, it will manifest itself as spalling and cracking of the surface of the concrete due to expansion because of the chemical reaction. Further, reaction with aggregates in inaccessible areas would also occur in accessible areas because aggregates were used in construction of both accessible and inaccessible areas. IWL, Examination Category L-A, requires periodic examination of accessible concrete surfaces and §50.55a(b)(2)(ix)(E) requires an evaluation of inaccessible areas when conditions exist in accessible areas that could indicate the presence of or result in degradation to such inaccessible areas. IWL and §50.55a(b)(2)(ix)(E) will detect such degradation.
	Expansion and cracking	

06	Concrete Structure Aging mechanism: Elevated temperature Aging effects: Loss of strength & modulus	Issue: NUREG-1557 states that elevated temperature is non-significant for concrete structures if it meets the following conditions: concrete containment structures be maintained at operating temperatures <66°C (150°F) and local area temperatures <93°C (200°F); or for concrete structures that experience temperatures greater than the above specified limits, a plant specific justification should be provided (Page B-32 of NUREG-1557). Recommendation: For concrete containment structures that experience temperatures greater than the above specified limits, a plant specific evaluation is needed. Discussion: Elevated temperature results in loss of concrete strength and modulus which may not be detected by the implementation of IWL and §50.55a modification until the aging effects are so severe as to result in cracking and spalling. Thus, for concrete structures that experience temperatures greater than the above specified limits, a plant specific justification should be provided.
07	Concrete Structure Aging mechanism: Irradiation of concrete Aging effects: Loss of strength & modulus	Issue: NUREG-1557 states that irradiation of concrete is non-significant for containment concrete structures (Page B-34 of NUREG-1557). Recommendation: The issue is non-significant. Discussion: The neutron fluence levels and maximum integrated gamma doses experienced by containment concrete during the license renewal term is not expected to exceed the level at which measurable degradation of concrete strength properties occurs (10^{19}n/cm^2 & 10^{10} rads, respectively). Thus the issue is non-significant.

08	Concrete Structure	Issue: NUREG-1557 states that concrete interaction with aluminum is non-significant for concrete containment structures if aluminum piping was not used for concrete placement, otherwise any adverse effects of concrete interactions with aluminum would have been identified during the initial acceptance test prior to initial operation (Page B-42 of NUREG-1557).
	Aging mechanism:	
	Concrete interaction with aluminum	
		Recommendation: Concrete interaction with aluminum is not an issue for license renewal.
	Aging effects:	
	Loss of strength	Discussion: Adverse effects of concrete interactions with aluminum would have occurred during the placement of concrete and would have been identified during initial structural acceptance test prior to plant initial operation. Identified concerns would have been evaluated during plant construction for appropriate corrective action. Any containment having concrete placed through aluminum pipelines which successfully completed its acceptance tests was not adversely affected by this placing condition. Thus it is not an issue for license renewal.

| 09 | Struct. Steel & Liner

Aging mechanism:

Corrosion

Aging effects:

Loss of material | Issue: NUREG-1557 indicates that corrosion is non-significant for above grade steel liner, steel containment shells (except at the proximity of the ice-condenser), and common steel components. NUREG-1557 indicates that galvanic corrosion is non-significant for penetration bellows if they are protected by shields from corrosive environment. NUREG-1557 indicates that inspection of structural steel and liner should be in accordance with IWE. NUREG-1557 also indicates that evaluation for management of inaccessible areas of below grade structural steel and liner is to be justified on a plant-specific basis (Pages B-37 and B-38 of NUREG-1557).

Recommendation: The issue would be managed with the implementation of IWE through §50.55a. However, the management of potential corrosion of inaccessible areas of structural steel liner, steel containment shells, and common steel components when conditions in accessible areas may not indicate the presence of or result in degradation to such inaccessible areas needs to be justified on a plant-specific basis.

Discussion: IWE, Examination Categories E-A, E-C, E-D, & E-G, provides periodic examination of accessible areas to uncover evidence of structural degradation and should detect corrosion of structural steel and liner; IWE, Examination Category E-P (Appendix J to 10 CFR 50, Type A test), requires a general inspection and an integrated leakage test; and §50.55a(b)(2)(x)(A) requires an evaluation of acceptability of inaccessible areas when conditions exist in accessible areas that could indicate the presence of or result in degradation to such inaccessible areas. However, the management of potential corrosion of inaccessible areas of structural steel liner, steel containment shells, and common steel components when conditions in accessible areas may not indicate the presence of or result in degradation to such inaccessible areas needs to be evaluated. |

10	Struct. Steel & Liner Aging mechanism: Elevated temperature Aging effects: Loss of strength & modulus	Issue: NUREG-1557 states that elevated temperature is non-significant for containment structural steel liner, steel containment shells, and common steel components such as penetration bellows/sleeves, personnel airlock, equipment hatches (Page B-33 of NUREG-1557). Recommendation: The issue is non-significant. Discussion: Operating temperatures within PWR containment structures are 49-66^{0}C (120-150^{0}F) which are well below the 316^{0}C (600^{0}F) level at which the structural integrity of rebar/concrete combination begins to be significantly affected. Thus the issue is non-significant.
11	Struct. Steel & Liner Aging mechanism: Irradiation of steel Aging effects: Loss of fracture toughness	Issue: NUREG-1557 states that irradiation of steel is non-significant for containment structural steel liner, containment shells, and common steel components (Page B-35 of NUREG-1557). Recommendation: The issue is non-significant. Discussion: The cumulative radiation exposure experienced by concrete containment liners or free-standing steel containment shells throughout the license renewal term is expected to be far below the level of 2×10^{17}n/cm^2(>1 MeV) which could cause a change in mechanical or physical properties. Thus the issue is non-significant.

| 12 | Struct. Steel & Liner

Aging mechanism:

Stress corrosion cracking (SCC)

Aging effects:

Crack initiation & growth | Issue: NUREG-1557 indicates that SCC is non-significant for concrete containment steel liner, free-standing steel containment shells, and common steel components in the containment environment unless dissimilar metal is used, and in the case of SS bellows assemblies for CS vent lines or pipe sleeves if the materials are protected by shields from corrosive environment (Page B-37 of NUREG-1557).

Recommendation: This issue would be managed by Examination Categories E-B & E-F of Subsection IWE and Appendix J to 10 CFR 50. In addition, an augmented VT-1 visual examination of bellows bodies should be performed using enhanced techniques qualified for detecting stress corrosion cracking in bellows bodies.

Discussion: IWE, Examination Category E-F, provides periodic surface examination of pressure retaining dissimilar metal welds for dissimilar metals and could detect SCC. IWE, Examination Category E-B, provides periodic visual examination of pressure retaining welds for containment penetrations. Also, any leakage associated with the containment shell or steel liner due to through-wall cracks resulting from SCC would be detected by periodic Appendix J leak rate test & remains within the limits of plant specifications or Subsection IWE. Although §50.55a indicates that Examination Categories E-B & E-F are optional during the current term of operation, these examinations should be performed for license renewal to demonstrate that no SCC has been initiated. In addition, since occurrences of transgranular stress corrosion cracking have been identified in operating plants on stainless steel bellows [Reference 15], an augmented examination on the surface areas of bellows bodies should be performed so that cracking would be detected. |

| 13 | <u>Reinforcing Steel</u>
<u>(Rebar)</u>

<u>Aging mechanism:</u>

Corrosion of
embedded steel

<u>Aging effects:</u>

Loss of bond & loss
of material | <u>Issue</u>: NUREG-1557 states that corrosion of embedded steel is non-significant for concrete structures above grade if not exposed to aggressive environment, pH<11.5 or chlorides >500 ppm; or if exposed to aggressive environment, concrete has relatively high strength [27.6 MPa (4 ksi)], low water-to-cement ratio (0.35-0.45), adequate air entrainment (3-6%), low permeability, and designed in accordance with ACI 318 or ASME Section III, Division 2. NUREG-1557 also indicates corrosion of embedded steel for concrete structures below grade exposed to aggressive ground water (pH <5.5, chloride >500 ppm, & sulfate >1500 ppm) should be examined in accordance with IWL and management of inaccessible areas should be justified on a case by case basis. Also the NRC staff considers that potential degradation due to chloride corrosion (e.g., ground water chemical attack) of PWR containments should be addressed. (Page B-36 of NUREG-1557).

<u>Recommendation</u>: The issue would be managed with the implementation of IWL through §50.55a. However, the management of potential corrosion of inaccessible areas of embedded steel when conditions in accessible areas may not indicate the presence of or result in degradation to such inaccessible areas needs to be justified on a plant-specific basis.

<u>Discussion</u>: IWL, Examination Category L-A, requires periodic examination of accessible concrete surfaces and §50.55a(b)(2)(ix)(E) requires an evaluation of inaccessible areas when conditions exist in accessible areas that could indicate the presence of or result in degradation to such inaccessible areas. Corrosion of embedded steel results in cracking and spalling of concrete and would be detected by inspections, regardless of whether the above conditions are met. However, the management of potential corrosion of inaccessible areas of embedded steel, when conditions in accessible areas may not indicate the presence of or result in degradation to such inaccessible areas needs to be evaluated. This would also address the staff's concern on chloride corrosion. |

14	Reinf. Steel & Prestr. Tendons	Issue: NUREG-1557 states that elevated temperature is non-significant for concrete containment reinforcing steel and for concrete containment prestressing tendons (Page B-32 of NUREG-1557).
	Aging mechanism:	Recommendation: The issue is non-significant, except for prestressed tendons. The tendon surveillance program should be augmented to include additional tendons based on their sun exposure or proximity to hot penetrations.
	Elevated temperature	
	Aging effects:	
	Loss of strength & modulus	Discussion: Operating temperatures within PWR containment structures are 49-66^{0}C (120-150^{0}F) which are well below the 316^{0}C (600^{0}F) level at which the structural integrity of rebar/concrete combination begins to be significantly affected. Thus the issue is non-significant for reinforcing steel. However, increase in temperature increases prestress loss in prestressed tendons. Prestress losses increased from 8% to 14% when the temperature was increased from 20^{0}C (68^{0}F) to 32^{0}C (90^{0}F) [Reference 16]. Thus, temperatures due to sun exposure or proximity to hot penetrations may increase the prestress loss in tendons. The tendon surveillance program described in Regulatory Guide 1.35 [Reference 10] is based on a small sample size, that is, a 4 percent random sample including a repeat tendon. Tendons subject to warm temperatures may not be tested because of this small sample size. The tendon surveillance program should be augmented to include additional tendons. These additional tendons should be selected based on their sun exposure or proximity to hot penetrations.

15	Reinf. Steel & Prestr. Tendons	Issue: NUREG-1557 states that irradiation of steel is non-significant for concrete structures reinforcing steel (including basemat reinforcing steel) and concrete containment prestressing tendons (Page B-34 of NUREG-1557).
	Aging mechanism: Irradiation of steel	Recommendation: The issue is non-significant.
	Aging effects: Loss of fracture toughness	Discussion: The cumulative radiation exposure experienced by reinforced concrete containment structures during the license renewal term is expected to be below the level of $10^{19} n/cm^2$ for degradation of reinforcing steel, and PWR concrete containment prestressing tendons & corrosion inhibitors will not receive enough radiation exposure during the license renewal term to incur age related degradation ($<4 \times 10^{16} n/cm^2$, & 10^{10} rads, respectively). Thus the issue is non-significant.

| 16 | Containment Structures & Components

Aging mechanism:

Fatigue

Aging effects:

Cumulative fatigue damage | Issue: NUREG-1557 states that fatigue is non-significant for containment structures and its components, except for the penetration sleeves and bellows. NUREG-1557 also indicates that fatigue is an unresolved issue for concrete containment penetration sleeves and steel containment penetration bellows and fatigue damage may not be detectable by a leak rate test (Pages B-40 & B-41 of NUREG-1557).

Recommendation: Fatigue is non-significant for containment structures and its components except for the penetration sleeves and bellows. Fatigue of containment penetration sleeves and penetration bellows is a "time-limited aging analysis" and must be evaluated in accordance with the license renewal rule, 10 CFR 54.21(c).

Discussion: Fatigue is non-significant for containment concrete, reinforcing steel, prestressing system components, steel liners, and free-standing steel containments, because they are designed to have good fatigue strength properties (10^5 cycles) of below yield load in accordance with ASME Section III, Division 2, or ACI 318, and ACI 215R-74 codes.

Containment penetration sleeves and penetration bellows are designed to Section III of the ASME Code which requires a fatigue analysis based on an assumed number of cycles. This fatigue analysis is a "time-limited aging analysis" and must be evaluated in accordance with license renewal rule §54.21(c) to ensure that the effects of aging on the intended functions will be adequately managed for the period of extended operation. |

17	Containment Structure & its Concrete basemat	Issue: NUREG-1557 states that settlement is an unresolved issue for containment concrete basemat for sites with soil, or significant changes in ground water conditions, and the effect of settlement needs to be evaluated (Page B-42 of NUREG-1557).
	Aging mechanism: Settlement Aging effects: Cracks, distortion, increase in component stress level	Recommendation: The issue is resolved by establishing a settlement monitoring program which would ensure that differential settlement of containment basemat does not exceed the design criteria for a containment structure and its basemat which is resting on soil or piles, or experiencing significant changes in ground water conditions. Discussion: Effects of differential settlement are potentially significant for a containment structure and its concrete basemat that is resting on soil or piles, or experiencing significant changes in ground water conditions. Subsection IWL does not address the effects of settlement. Because the effects of settlement could cause cracks and distortion of concrete basemat and could result in increasing stress levels greater than original design basis in the basemat and other parts of the containment structure. The effects of settlement of PWR containments need to be evaluated. However, a settlement monitoring program could ensure that the differential settlement does not exceed the design criteria for the containment structures throughout the license renewal term. A settlement monitoring program should be provided to manage settlement for containment basemat bearing on soil or piles, or experiencing significant changes in ground water conditions for the period of extended operation.

| 18 | Containment Structure & its Concrete basemat

Aging mechanism:

Erosion of cement

Aging effects:

Loss of strength | Issue: NRC Information Notice 97-11 indicates that erosion of cement from porous concrete could be potentially significant for porous concrete subfoundations below concrete basemat if subfoundation layers of porous concrete are used in the construction of concrete basemat with the presence of underground water.

Recommendation: For those applicable plants, the management of potential erosion of cement from porous concrete needs to be justified on a plant-specific basis.

Discussion: §50.55a(b)(2)(ix)(E) requires an evaluation of inaccessible areas when conditions exist in accessible areas that could indicate the presence of or result in degradation to such inaccessible areas. However, the management of potential erosion of cement from porous concrete of inaccessible areas of containment concrete basemat when conditions in accessible areas may not indicate the presence of or result in degradation to such inaccessible areas needs to be evaluated on a plant-specific basis. |

| 19 | Containment Structure & its Components | Issue: NUREG-1557 indicates that strain aging is non-significant for steel containment structures (including common components, such as penetration sleeves, penetration bellows, personnel airlock, and equipment hatches) that meet the following conditions: Dynamic strain aging is non-significant for free standing steel containment structures that do not allow loads to exceed the elastic limit. Static strain aging is non-significant for free standing steel containment structures that were not cold worked; or if cold worked during the forming process, the plates were normalized or stress relieved or both after forming with minimal (<5%) subsequent cold working (Page B-43 of NUREG-1557). |
| | Aging mechanism: Strain aging of carbon steel Aging effects: Loss of fracture toughness | Recommendation: This issue is non-significant. Discussion: Dynamic strain aging is not expected in the carbon steel components of steel containments during their service life, since the strains associated with the design service loads are below the elastic limit of the material. The PWR containment is made from low-carbon steel, and the steel is normalized or stress relieved or both following plate rolling. Further, strain aging requires stressing of the material to above its yield stress, and aging at temperatures above 93^0C (200^0F). Carbon-related strain aging at temperatures below 93^0C (200^0F) is normally negligible due to the low solubility of carbon in this temperature range. The PWR containment has a maximum temperature during normal operation of about 66^0C (150^0F), and loading conditions do not produce service stresses in the range of the material yield strength. Thus strain aging is non-significant for the steel containment structures and its components. |

| 20 | Conc. Containment Prestr. Tendons

Aging mechanism:

Stress relaxation of prestressing wire, shrinkage creep, anchorage seating losses, and tendon friction

Aging effects:

Loss of prestress | Issue: NUREG-1557 indicates that loss of prestress due to stress relaxation, shrinkage creep, etc., would be a reduction of design margin and could be potentially significant for prestressing tendons for license renewal (Page B-41 of NUREG-1557).

Recommendation: Loss of prestress for prestressing tendons would be managed with the implementation of Subsection IWL through §50.55a. In addition, the tendon prestress evaluation is a "time-limited aging analysis" and must be evaluated in accordance with the license renewal rule, 10 CFR 54.21(c).

Discussion: Subsection IWL, Examination Category L-B, and §50.55a inspections would be able to detect potential loss of prestress for prestressing tendons. For example: IWL-2522 provides examination method for tendon force measurements; IWL-3221 provides acceptance standard for measuring tendon force; §50.55a(b)(2)(ix)(B) states that "when evaluation of consecutive surveillances of prestressing forces for the same tendon ... indicates a trend of prestress loss ...., an evaluation shall be performed;" repair and replacement are addressed in IWL-4000 and IWL-7000, respectively. In addition, the recommendations of Regulatory Guide 1.35 [Reference 10] were incorporated into the 1992 Edition with the 1992 Addenda of Subsection IWL and §50.55a(b)(2)(ix)(A)-(D), the final rule, dated August 8, 1996 (61 FR 41304). Thus the "loss of prestress" for prestressing tendons would be managed with the implementation of IWL through §50.55a. Further, the tendon prestress evaluation is a "time-limited aging analysis" and must be evaluated for renewal to demonstrate that the prestressing force will meet the design requirements at the end of 60 years in accordance with license renewal rule §54.21(c). |

21	Conc. Containment Prestr. Tendons	Issue: NUREG-1557 states that corrosion of tendons is an unresolved issue in that the NRC staff is concerned that a large amount of grease leakage can degrade concrete strength. IWL (1992 Edition with 1992 Addenda) lacks certain criteria contained in RG 1.35. These criteria are addressed in 10 CFR 50.55a final rule, dated August 8, 1996, Section 50.55a(b)(2)(ix)(A)-(D) on issues such as failed wires and tendon sheathing filler grease conditions. Also, anchor heads have failed in prestressed concrete containments. NUREG-1557 also states that prestressing tendons and tendon anchorage hardware should be examined in accordance with the provisions of RG 1.35 for prestressed concrete containments (Page B-39 of NUREG-1557).
	Aging mechanism: Corrosion of tendons Aging effects: Loss of material	
		Recommendation: The issue is resolved with the implementation of Subsection IWL through §50.55a.
		Discussion: Subsection IWL, Examination Category L-B, and §50.55a inspections would be able to detect corrosion of prestressing tendons. For example: IWL-2525 provides methods for examination of corrosion protection medium and free water; IWL-3221 provides acceptance standard for corrosion protection medium; IWL-2524 provides visual examination of tendon anchorage areas; IWL-3221 provides acceptance standard for inspection of tendon anchorage areas; §50.55a(b)(2)(ix)(A) states that "grease caps that are accessible must be visually examined to detect grease leakage or grease cap deformations....... that indicates deterioration of anchorage hardware." In addition, the recommendations of Regulatory Guide 1.35 [Reference 10] were incorporated into the 1992 Edition with the 1992 Addenda of Subsection IWL and §50.55a(b)(2)(ix)(A)-(D), the final rule, dated August 8, 1996 (61 FR 41304). Thus the "Corrosion of prestressing tendons" would be managed with the implementation of IWL through §50.55a.

| 22 | Containment Pressure Retaining Components

Aging mechanism:

Mechanical Wear

Aging effects:

Fretting, Lockup | Issue: NUREG-1557 does not address mechanical wear for PWR containments. However, mechanical wear could be potentially significant for components that are subject to relative sliding or rotating motion and that are susceptible to fretting and/or lockup.

Recommendation: Mechanical wear for PWR containment pressure retaining components would be managed with the implementation of Subsection IWE (Examination Categories E-D, E-G, and E-P).

Discussion: NUREG-1557 addresses mechanical wear issue only for BWR containments. However, mechanical wear should also be addressed for PWR containments. Inspection and mitigation of mechanical wear conducted in accordance with the provisions of Subsections IWE & IWF, as applicable, would ensure that the integrity of containment pressure retaining components and their supports is maintained throughout the license renewal term. IWE, Examination Categories E-D & E-G, provides periodic examinations for the pressure retaining components (including airlocks, and equipment hatches). IWE, Examination Category E-P (Appendix J to Part 50, Type B test), would detect local leaks for those components. Thus any potential mechanical wear degradation would be detected by the implementation of IWE for those pressure retaining components. (There are no PWR containment pressure retaining component supports listed in Appendix B of this report and therefore, Subsection IWF is not applicable.) |

*   Evaluation is based on:
(1) the 1992 Edition with the 1992 Addenda of Subsections IWE and IWL of Section XI of the ASME Code;
(2) the 1989 Edition of Section XI of the ASME Code including Appendix VII, "Qualification of Nondestructive Examination Personnel for Ultrasonic Examination," and Appendix VIII (1989 Addenda), "Performance Demonstration for Ultrasonic Examination Systems;"
(3) the final rule on 10 CFR 50.55a, Codes and Standards, dated August 8, 1996 (61 FR 41303); and
(4) NUREG-1557, "Summary of Technical Information and Agreements from Nuclear Management and Resources Council Industry Reports Addressing License Renewal," dated October 1996.

TABLE 2.   AGING MANAGEMENT OF BWR CONTAINMENTS FOR LICENSE RENEWAL.

Item	Component, Aging Mechanism & Aging Effects	Issue and Evaluation*
01	<u>Concrete & Steel Containment</u> <u>Aging mechanism:</u> Not applicable. <u>Aging effects:</u> General	<u>Issue:</u> A "one-time inspection for license renewal." NUREG-1557 states that one time inspection is an unresolved issue regarding staff request for inspection of concrete containment & steel containment to assess the current condition of containment and to provide a baseline information for any future inspections (Page B-28 of NUREG-1557). <u>Recommendation:</u> The issue is resolved with the implementation of IWE/IWL through §50.55a. However, specific-recommendations for applicable aging effects are addressed hereinafter in this table. <u>Discussion:</u> Subsections IWE and IWL require periodic inspection of the containment in accessible areas. These inspections would periodically assess the condition of the containment and each inspection would provide a documented baseline for subsequent inspections. Furthermore, §50.55a(b)(2)(ix)(E) and (b)(2)(x)(A) require an evaluation of inaccessible areas when conditions exist in accessible areas that could indicate the presence of or result in degradation to such inaccessible areas. However, the management of potential aging effects of inaccessible areas, when conditions in accessible areas may not indicate the presence of or result in degradation to such inaccessible areas, is addressed individually for each applicable aging effect (i.e., Items 1, 3, 4, 9, and 13 of this table). Conditions for such inaccessible areas should be evaluated for license renewal. A program for a one-time inspection may be proposed.

NUREG-1611

| 02 | Concrete Structure

Aging mechanism:

Freeze-thaw

Aging effects:

Scaling, cracking, & spalling | Issue: NUREG-1557 states that freeze-thaw is non-significant for Mark I & Mark II concrete containments because they are protected from freezing by the secondary containment. Freeze-thaw is non-significant for Mark III concrete containment components if the following conditions are met: Mark III concrete containment components located in a geographic regions of negligible weathering conditions (weathering index <100 day-inch/yr); and if located in severe (weathering index >500 day-inch/yr) or moderate (100-500 day-inch/yr) weathering conditions with the concrete mix design meets the air content & water-to-cement ratio requirements of ASTM C260-77; or equivalently, the ASME Sect. III, Division 2, paragraph CC 2231.7.1; or the susceptible surfaces are protected by shielding (Page B-46 of NUREG-1557).

Recommendation: The issue is resolved with the implementation of IWL.

Discussion: Freeze-thaw results in scaling, cracking, and spalling. Any freeze-thaw degradation would initially appear in the exposed concrete structure. Subsection IWL, Examination Category L-A, requires periodic visual examination of accessible concrete surfaces and would detect any freeze-thaw damage of the concrete containment, including the dome, regardless of whether the above weathering conditions are met. |

| 03 | Concrete Structure

Aging mechanism:

Leaching of calcium hydroxide

Aging effects:

Increase of porosity & permeability | Issue: NUREG-1557 states that leaching of calcium hydroxide is non-significant for containment concrete structures if the following conditions are met: concrete structures not exposed to flowing water; and for concrete structures that are exposed to flowing water but are constructed using the guidance of ACI 201.2R-77 to ensure dense, well-cured concrete with low permeability and control cracking through proper arrangement and distribution of reinforcement (Page B-47 of NUREG-1557).

Recommendation: The issue would be managed with the implementation of IWL through §50.55a. However, the management of potential leaching of calcium hydroxide of inaccessible areas of containment concrete structures when conditions in accessible areas may not indicate the presence of or result in degradation to such inaccessible areas needs to be justified on a plant-specific basis.

Discussion: IWL, Examination Category L-A, requires periodic examination of accessible concrete surfaces and §50.55a(b)(2)(ix)(E) requires an evaluation of inaccessible areas when conditions exist in accessible areas that could indicate the presence of or result in degradation to such inaccessible areas. Regardless of whether the above conditions are met, potential leaching of calcium hydroxide would be detected as water stains on accessible surfaces by the IWL visual examination. However, the management of potential leaching of calcium hydroxide of inaccessible areas (e.g., below grade portion of concrete structures with presence of flowing water) when conditions in accessible areas may not indicate the presence of or result in degradation to such inaccessible areas, needs to be evaluated on a plant-specific basis. |

| 04 | Concrete Structure Aging mechanism: Aggressive chemical attack Aging effects: Increase of porosity and permeability, cracking, and spalling | Issue: NUREG-1557 states that aggressive chemical attack is non-significant for above grade containment concrete structures because they are not exposed to ground water. Aggressive chemical attack is non-significant for below grade containment concrete structures if the following conditions are met: containment concrete is not exposed to aggressive ground water (pH <5.5, chloride >500 ppm, & sulfate >1500 ppm); or if expos%d to ground water that exceeds the pH, chloride, sulfate limits, the exposure is for intermittent periods only. NUREG-1557 indicates that inspection of containment concrete structure should be in accordance with IWL. NUREG-1557 states that evaluation for management of inaccessible areas of below grade containment concrete structures is to be justified on a plant-specific basis (Page B-48 of NUREG-1557). Recommendation: The issue would be managed with the implementation of IWL through §50.55a. However, the management of potential aggressive chemical attack of inaccessible areas of containment concrete structures when conditions in accessible areas may not indicate the presence of or result in degradation to such inaccessible areas needs to be justified on a plant-specific basis. Discussion: Aggressive chemical attack results in increase of porosity and permeability, cracking and spalling. IWL, Examination Category L-A, requires periodic examination of accessible concrete surfaces and §50.55a(b)(2)(ix)(E) requires an evaluation of inaccessible areas when conditions exist in accessible areas that could indicate the presence of or result in degradation to such inaccessible areas. Regardless of whether the above conditions are met, potential aggressive chemical attack would be detected by IWL and §50.55a(b)(2)(ix)(E). However, the management of potential aggressive chemical attack of inaccessible areas when conditions in accessible areas may not indicate the presence of or result in degradation to such inaccessible areas needs to be evaluated. |

| 05 | Concrete Structure

Aging mechanism:

Reaction with aggregates

Aging effects:

Expansion and cracking | Issue: NUREG-1557 states that reaction with aggregates is an unresolved issue. NUREG-1557 indicates that the NRC staff believes that alkaline-aggregate reactions can not be ruled out. Tests involving aggregates alone are not satisfactory in predicting aggregate performance. Alkaline-aggregate reaction may occur after 25 or more years (Page B-49 of NUREG-1557).

Recommendation: The issue is resolved with the implementation of IWL through §50.55a.

Discussion: If alkaline-aggregate reaction occurs, it will manifest itself as spalling and cracking of the surface of the concrete due to expansion because of the chemical reaction. Further, reaction with aggregates in inaccessible areas would also occur in accessible areas because aggregates were used in construction of both accessible and inaccessible areas. IWL, Examination Category L-A, requires periodic examination of accessible concrete surfaces and §50.55a(b)(2)(ix)(E) requires an evaluation of inaccessible areas when conditions exist in accessible areas that could indicate the presence of or result in degradation to such inaccessible areas. IWL and §50.55a(b)(2)(ix)(E) will detect such degradation. |

06	Concrete Structure Aging mechanism: Elevated temperature Aging effects: Loss of strength & modulus	Issue: NUREG-1557 states that elevated temperature is non-significant for concrete structures if it meets the following conditions: concrete containment structures be maintained at operating temperatures <66°C (150°F) and local area temperatures <93°C (200°F); or for concrete structures that experience temperatures greater than the above specified limits, a plant specific justification should be provided (Page B-56 of NUREG-1557). Recommendation: For concrete structures that experience temperatures greater than the above specified limits, a plant specific evaluation should be performed. Discussion: Elevated temperature results in loss of concrete strength and modulus which may not be detected by the implementation of IWL and §50.55a modification until the aging effects are so severe as to result in cracking and spalling. Thus, for concrete structures that experience temperatures greater than the above specified limits, a plant specific justification should be provided.
07	Concrete Structure Aging mechanism: Irradiation of concrete Aging effects: Loss of strength & modulus	Issue: NUREG-1557 states that irradiation of concrete is non-significant for containment concrete structures (Page B-57 of NUREG-1557). Recommendation: The issue is non-significant. Discussion: The neutron fluence levels and maximum integrated gamma doses experienced by containment concrete during the license renewal term is not expected to exceed the level at which measurable degradation of concrete strength properties occurs (10^{19} n/cm^2 neutron radiation & 10^{10} rads gamma radiation, respectively for concrete). Thus the issue is non-significant.

| 08 | Struct. Steel & Liner

Aging mechanism:

Atmospheric corrosion

Aging effects:

Loss of material | Issue: NUREG-1557 states that atmospheric corrosion is non-significant for containment steel components if the following conditions are met: (a) containment steel components fabricated from stainless steel, or for components having intact protective coating, or for components having a corrosion allowance $\geq 1/32$ inch. Austenitic SS is corrosion resistant. The atmospheric corrosion for carbon and low alloy steels without protective coatings is less than 0.5 mils per year or <1/32 inches for a 60-year period, and (b) the examination categories E-A, E-P, & E-C of ASME Sect. XI, Subsect. IWE are required to be performed in conjunction with 10 CFR 50, Appendix J, Type A leak rate test (Pages B-50 & 51 of NUREG-1557).

Recommendation: This issue would be managed with the implementation of IWE through §50.55a.

Discussion: IWE, Examination Categories E-A & E-C, requires periodic examination of accessible surfaces for containment steel structures & its components; Examination Category E-P (Appendix J to 10 CFR 50, Type A test), requires a general inspection and an integrated leakage test; and §50.55a(b)(2)(x)(A) requires an evaluation of inaccessible areas when conditions exist in accessible areas that could indicate the presence of or result in degradation to such inaccessible areas. Atmospheric corrosion may exist when relevant conditions for coated areas including evidence of flaking, blistering, peeling, discoloration, etc., and relevant conditions for uncoated areas including evidence of cracking, discoloration, wear, pitting, excessive corrosion, etc., occur. If the examination areas are found to be defective, or to be suspectable, the augmented examinations of IWE-1240 will apply to ensure that minimum wall thickness is properly evaluated and maintained in accordance with IWE-3512. |

| 09 | Struct. Steel & Liner

Aging mechanism:

Local corrosion

Aging effects:

Loss of material | Issue: (a) NUREG-1557 indicates that local corrosion would be managed with the implementation of IWE-1240 for steel containment and its common components. NUREG-1557 indicates that a plant-specific aging program is required to manage the local corrosion of steel containment inaccessible areas and/or embedded carbon steel containment components (Pages B-53 and B-55 of NUREG-1557). (b) NUREG-1557 indicates that local corrosion is non-significant for concrete containment liners and anchors if the following conditions are met: Corrosion of the liner plate is mitigated by protective coatings on the interior surface, and by the alkaline environment between the exterior surface of the liner plate and the concrete structure. Stainless steel is corrosion resistant (Pages B-54 of NUREG-1557).

Recommendation: All accessible areas would be managed with implementing IWE through §50.55a. However, the management of potential local corrosion of inaccessible areas of structural steel and liner when conditions in accessible areas may not indicate the presence of or result in degradation to such inaccessible areas needs to be justified on a plant-specific basis.

Discussion: IWE, Examination Categories E-A, E-C, E-D, and E-G, provides periodic examination of accessible areas to uncover structural degradation. This inspection would detect local corrosion regardless whether the conditions in (b) are met. IWE-1240 specifies augmented inspections for areas likely to experience accelerated degradation and aging. §50.55a(b)(2)(x)(A) requires an evaluation of acceptability of inaccessible areas when conditions exist in accessible areas that could indicate the presence of or result in degradation to such inaccessible areas. However, the management of local corrosion of inaccessible areas when conditions in accessible areas may not indicate the presence of or result in degradation to such inaccessible areas needs to be evaluated. |

| 10 | Struct. Steel & Liner

Aging mechanism:

Elevated temperature

Aging effects:

Loss of strength & modulus | Issue: NUREG-1557 states that elevated temperature is non-significant for PWR containment structural steel and liner (Page B-33 of NUREG-1557). However, NUREG-1557 does not address elevated temperature effects on BWR containment steel liner (Page B-56 of NUREG-1557).

Recommendation: The issue is non-significant.

Discussion: Operating temperatures within BWR containment structures are $<66^{0}C$ ($150^{0}F$) which are well below the $316^{0}C$ ($600^{0}F$) level at which the structural integrity of rebar/concrete combination begins to be significantly affected. Thus the issue is non-significant. This conclusion is also applicable to the BWR containment steel liner. |
| 11 | Struct. Steel & Liner

Aging mechanism:

Irradiation of steel

Aging effects:

Loss of fracture toughness | Issue: NUREG-1557 states that irradiation of steel is non-significant for containment structural steel & liner (Page B-57 of NUREG-1557).

Recommendation: The issue is non-significant.

Discussion: The neutron fluence levels & maximum integrated gamma doses incurred by containment components, including containment steel & liners throughout the license renewal period are not expected to exceed the level at which measurable degradation occurs ($2x10^{17}n/cm^{2}$ for all components made of carbon steel, stainless steel, and liner plate). Thus the issue is non-significant. |

12	Struct. Steel & Liner	Issue: (a) NUREG-1557 indicates that SCC is non-significant for containment components, including penetration sleeves, bellows, and vent line bellows if the following conditions are met: for austenitic SS containment components that are only exposed to the containment or reactor building environment or their normal operational stress levels are less than materials yield strength or fracture mechanics analysis has established that cracks do not propagate; and for high strength bolts if material yield strength is <1034 MPa(<150 ksi). (b) SCC would be managed for suppression chamber shell interior surface by implementing 10 CFR 50, Appendix J integrated leak rate test to maintain liner integrity (Page B-63 of NUREG-1557).
	Aging mechanism:	
	Stress corrosion cracking (SCC)	
	Aging effects:	
	Crack initiation & growth	

Recommendation: This issue would be managed by Examination Categories E-B & E-F of Subsection IWE and Appendix J to 10 CFR 50. In addition, an augmented VT-1 visual examination of bellows bodies should be performed using enhanced techniques qualified for detecting stress corrosion cracking in bellows bodies.

Discussion: IWE Examination Category E-F provides periodic surface examination of pressure retaining dissimilar metal welds for dissimilar metals and could detect SCC. IWE, Examination Category E-B, provides periodic visual examination of pressure retaining welds for containment penetrations. Also any leakage associated with the steel liner, including suppression pool liner, due to through-wall cracks resulting from SCC would be detected by periodic Appendix J leak rate test & remains within the limits of plant specifications or Subsection IWE. Although §50.55a indicates that Examination Categories E-B & E-F are optional during the current term of operation, these examinations should be performed for license renewal to demonstrate that no SCC has been initiated. In addition, since occurrences of transgranular stress corrosion cracking have been identified in operating plants on SS bellows [Reference 15], an augmented examination on the surface areas of bellows bodies should be performed so that cracking would be detected.

13	Reinforcing Steel (Rebar) Aging mechanism: Corrosion of embedded steel Aging effects: Loss of bond & loss of material	Issue: NUREG-1557 states that corrosion of embedded steel is non-significant for concrete structures not exposed to aggressive environment (pH<11.5 or chlorides >500 ppm); or for concrete exposed to aggressive environment but has relatively high strength [27.6 MPa (4 ksi)] and low water-to-cement ratio (0.35-0.45), adequate air entrainment (3-6%), low permeability, and are designed in accordance with ACI 318 or ASME Section III, Division 2. NUREG-1557 indicates corrosion of embedded steel for concrete structures below grade exposed to aggressive ground water (pH <5.5, chloride >500 ppm, & sulfate >1500 ppm) should be examined in accordance with IWL and management of inaccessible areas should be justified on a case by case basis. Also the NRC staff considers that potential degradation due to chloride corrosion (e.g., ground water chemical attack) of containments should be addressed (Page B-52 of NUREG-1557). Recommendation: The issue would be managed with the implementation of IWL through §50.55a. However, the management of potential corrosion of inaccessible areas of embedded steel when conditions in accessible areas may not indicate the presence of or result in degradation to such inaccessible areas needs to be justified on a plant-specific basis. Discussion: IWL, Examination Category L-A, requires periodic examination of accessible areas and §50.55a(b)(2)(ix)(E) requires an evaluation of inaccessible areas when conditions exist in accessible areas that could indicate the presence of or result in degradation to such inaccessible areas. Corrosion of embedded steel results in cracking and spalling of concrete and would be detected by inspections, regardless of whether the above conditions are met. However, the management of potential corrosion of inaccessible areas of embedded steel, when conditions in accessible areas may not indicate the presence of or result in degradation to such inaccessible areas needs to be evaluated. This would also address the staff's concern on chloride corrosion.

14	Reinf. Steel & Prestr. Tendons	Issue: NUREG-1557 states that elevated temperature is non-significant for concrete containment reinforcing steel and for concrete containment prestressing tendons (Page B-56 of NUREG-1557).
	Aging mechanism:	
	Elevated temperature	Recommendation: The issue is non-significant, except for prestressed tendons. The tendon surveillance program should be augmented to include additional tendons based on their sun exposure or proximity to hot penetrations.
	Aging effects:	
	Loss of strength & modulus	Discussion: Operating temperatures within BWR containment structures are <66°C(150°F) which are well below the 316°C (600°F) level at which the structural integrity of rebar/concrete combination begins to be significantly affected. Additionally, concrete containment prestressing tendons are normally subjected to temperatures <60°C (140°F). Thus the issue is non-significant for reinforcing steel. However, increase in temperature increases prestress loss in prestressed tendons. Prestress losses increased from 8% to 14% when the temperature was increased from 20°C (68°F) to 32°C (90°F) [Reference 16]. Thus, temperatures due to sun exposure or proximity to hot penetrations may increase the prestress loss in tendons. The tendon surveillance program described in Regulatory Guide 1.35 [Reference 10] is based on a small sample size, that is, a 4 percent random sample including a repeat tendon. Tendons subject to warm temperatures may not be tested because of this small sample size. The tendon surveillance program should be augmented to include additional tendons. These additional tendons should be selected based on their sun exposure or proximity to hot penetrations.

15	Reinf. Steel & Prestr. Tendons	Issue: NUREG-1557 states that irradiation of steel is non-significant for concrete structures reinforcing steel (including basemat reinforcing steel) and concrete containment prestressing tendons (Page B-57 of NUREG-1557).
	Aging mechanism:	Recommendation: The issue is non-significant.
	Irradiation of steel	
		Discussion: The neutron fluence levels & maximum integrated gamma doses incurred by containment components, including rebars & prestressed tendons for both the current and license renewal period are not expected to exceed the level at which measurable degradation occurs. (4×10^{16} n/cm^2 for concrete containment prestressing tendons; 2×10^{17} n/cm^2 for all components made of CS, SS including rebar, and liner). Thus the issue is non-significant.
	Aging effects:	
	Loss of fracture toughness	

16	Containment Structures & Components	Issue: NUREG-1557 states that fatigue is non-significant for containment structures and its components, except for the penetration sleeves and bellows. NUREG-1557 also indicates that fatigue is an unresolved issue for containment penetration sleeves and penetration bellows and fatigue damage may not be detectable by a leak rate test (Pages B-58 & B-59 of NUREG-1557).
	Aging mechanism: Fatigue	Recommendation: Fatigue is non-significant for containment structures and its components except for the penetration sleeves and bellows. Fatigue of containment penetration sleeves and penetration bellows is a "time-limited aging analysis" and must be evaluated in accordance with license renewal rule §54.21(c).
	Aging effects: Cumulative fatigue damage.	Discussion: Fatigue is non-significant for containment concrete, reinforcing steel, prestressing system components, steel liners, and free-standing steel containments, because they are designed to have good fatigue strength properties (10^5 cycles) of below yield load in accordance with ASME Section III, Division 2, or ACI-318, and ACI 215R-74 codes.
		Containment penetration sleeves and penetration bellows are designed to Section III of the ASME Code which requires a fatigue analysis based on an assumed number of cycles. This fatigue analysis is a "time-limited aging analysis" and must be evaluated in accordance with license renewal rule §54.21(c) to ensure that the effects of aging on the intended functions will be adequately managed for the period of extended operation.

| 17 | Containment Structure & its Concrete Basemat

Aging mechanism:

Settlement

Aging effects:

Cracks, distortion, increase in component stress level | Issue: NUREG-1557 indicates that for BWR containment concrete basemat bearing on soil or piles, or experiencing significant changes in ground water conditions, a settlement monitoring program is required to ensure that the differential settlement does not exceed the design criteria for the containment throughout the license renewal term (Page B-62 of NUREG-1557).

Recommendation: The issue would be managed by establishing a settlement monitoring program which would ensure that differential settlement of containment basemat does not exceed the design criteria for a containment structure and its basemat bearing on soil or piles, or experiencing significant changes in ground water conditions.

Discussion: Effects of differential settlement are potentially significant for a containment structure and its concrete basemat that is resting on soil or piles, or experiencing significant changes in ground water conditions. Subsection IWL does not address the effects of settlement. Because the effects of settlement could cause cracks and distortion of concrete basemat and could result in increasing stress levels greater than the original design basis in the basemat and other parts of the containment structure. A settlement monitoring program could ensure that the differential settlement does not exceed the design criteria for the containment structures throughout the license renewal term. A settlement monitoring program should be provided to manage settlement for containment basemat bearing on soil or piles, or experiencing significant changes in ground water conditions for the period of extended operation. |

18	Containment Structure & its Concrete basemat	Issue: NRC Information Notice 97-11 indicates that erosion of cement from porous concrete could be potentially significant for porous concrete subfoundations below concrete basemat if subfoundation layers of porous concrete are used in the construction of concrete basemat with the presence of underground water.
	Aging mechanism:	Recommendation: For those applicable plants, the management of potential erosion of cement from porous concrete needs to be justified on a plant-specific basis.
	Erosion of cement	
	Aging effects:	Discussion: §50.55a(b)(2)(ix)(E) requires an evaluation of inaccessible areas when conditions exist in accessible areas that could indicate the presence of or result in degradation to such inaccessible areas. However, the management of potential erosion of cement from porous concrete of inaccessible areas of containment concrete basemat, when conditions in accessible areas may not indicate the presence of or result in degradation to such inaccessible areas, needs to be evaluated on a plant-specific basis.
	Loss of strength	

| 19 | Containment Structure & its Components

Aging mechanism:

Strain aging of carbon steel

Aging effects:

Loss of fracture toughness | Issue: NUREG-1557 indicates that strain aging is non-significant for steel containment structures (including common components, such as penetration sleeves, penetration bellows, personnel airlock, equipment hatches, and CRD hatch), and concrete containment steel components (including vent lines, vent line bellows, and drywell head) that meet the following conditions: Dynamic strain aging is non-significant for containment steel components that do not allow loads to exceed the elastic limit. Static strain aging is non-significant for containment steel components that were not cold worked; or if cold worked during the forming process, the plates were normalized or stress relieved or both after forming with minimal (<5%) subsequent cold working (Page B-61 of NUREG-1557).

Recommendation: This issue is non-significant.

Discussion: Dynamic strain aging is not expected in the carbon steel components of containments during their service life, since the strains associated with the design service loads are below the elastic limit of the material. The BWR containment is made from low-carbon steel, and the steel is normalized or stress relieved or both following plate rolling. Further, strain aging requires stressing of the material to above its yield stress, and aging at temperatures above 93^0C (200^0F). Carbon-related strain aging at temperatures below 93^0C (200^0F) is normally negligible due to the low solubility of carbon in this temperature range. The BWR containment has a maximum temperature during normal operation of about 66^0C (150^0F), and loading conditions do not produce service stresses in the range of the material yield strength. Thus strain aging is non-significant for containment steel structure and its steel components. |

| 20 | Conc. Containment Prestr. Tendons Aging mechanism: Stress relaxation of prestressing wire, shrinkage creep, anchorage seating losses, and tendon friction Aging effects: Loss of prestress | Issue: NUREG-1557 indicates that loss of prestress due to stress relaxation, shrinkage creep, etc., would be a reduction of design margin and could be potentially significant for prestressing tendons for license renewal (Page B-62 of NUREG-1557). Recommendation: Loss of prestress for prestressing tendons would be managed with the implementation of Subsection IWL through §50.55a. In addition, the tendon prestress evaluation is a "time-limited aging analysis" and must be evaluated in accordance with the license renewal rule, 10 CFR §54.21(c). Discussion: Subsection IWL, Examination Category L-B, and §50.55a inspections would be able to detect potential loss of prestress for prestressing tendons. For example: IWL-2522 provides examination method for tendon force measurements; IWL-3221 provides acceptance standard for measuring tendon force; §50.55a(b)(2)(ix)(B) states that "when evaluation of consecutive surveillances of prestressing forces for the same tendon ... indicates a trend of prestress loss ...., an evaluation shall be performed;" repair and replacement are addressed in IWL-4000 and IWL-7000, respectively. In addition, the recommendations of Regulatory Guide 1.35 [Reference 10] were incorporated into the 1992 Edition with the 1992 Addenda of Subsection IWL and §50.55a(b)(2)(ix)(A)-(D), the final rule, dated August 8, 1996 (61 FR 41304). Thus the "loss of prestress" for prestressing tendons would be managed with the implementation of IWL through §50.55a. Further, the tendon prestress evaluation is a "time-limited aging analysis" and must be evaluated for renewal to demonstrate that the prestressing force will meet the design requirements at the end of 60 years in accordance with license renewal rule §54.21(c). |

| 21 | Conc. Containment Prestr. Tendons

Aging mechanism:

Corrosion of tendons

Aging effects:

Loss of material | Issue: NUREG-1557 states that corrosion of tendons is an unresolved issue in that the NRC staff is concerned that a large amount of grease leakage can degrade concrete strength. IWL (1992 Edition with 1992 Addenda) lacks certain criteria contained in RG 1.35. These criteria are addressed in 10 CFR 50.55a final rule, dated August 8, 1996, Section 50.55a(b)(2)(ix)(A)-(D) on issues such as failed wires and tendon sheathing filler grease conditions. Also, anchor heads have failed in prestressed concrete containments. NUREG-1557 states that prestressing tendons and tendon anchorage hardware should be examined in accordance with the provisions of RG 1.35 for prestressed concrete containments (Page B-55 of NUREG-1557).

Recommendation: The issue is resolved with the implementation of Subsection IWL through §50.55a.

Discussion: Subsection IWL, Examination Category L-B, and §50.55a inspections would be able to detect corrosion of prestressing tendons. For example: IWL-2525 provides methods for examination of corrosion protection medium and free water; IWL-3221 provides acceptance standard for corrosion protection medium; IWL-2524 provides visual examination of tendon anchorage areas; IWL-3221 provides acceptance standard for inspection of tendon anchorage areas; §50.55a(b)(2)(ix)(A) states that "grease caps that are accessible must be visually examined to detect grease leakage or grease cap deformations....... that indicates deterioration of anchorage hardware." In addition, the recommendations of Regulatory Guide 1.35 [Reference 10] were incorporated into the 1992 Edition with the 1992 Addenda of Subsection IWL and §50.55a(b)(2)(ix)(A)-(D), the final rule, dated August 8, 1996 (61 FR 41304). Thus the "Corrosion of prestressing tendons" would be managed with the implementation of IWL through §50.55a. |

22	Containment Pressure Retaining Components Aging mechanism: Mechanical wear Aging effects: Fretting, Lockup	Issue: Mechanical wear could be potentially significant for components that are subject to relative sliding or rotating motion and that are susceptible to fretting and/or lockup. (Page B-60 of NUREG-1557). Recommendation: Mechanical wear for BWR containment pressure retaining components & their supports would be managed with the implementation of Subsections IWE and IWF. Discussion: Inspection and mitigation of mechanical wear conducted in accordance with the provisions of Subsections IWE & IWF would ensure that the integrity of containment pressure retaining components and their supports is maintained throughout the license renewal term. IWE, Examination Categories E-D & E-G, provides periodic examinations for the pressure retaining components, (including airlock, equipment hatch, CRD hatch & drywell head). IWE, Examination Category E-P (Appendix J to Part 50, Type B test), would detect local leaks for those components. The supporting components such as downcomer bracing, column & saddle supports, seismic restraints & vent system supports are considered MC component supports which are periodically examined by Examination Category F-A of Subsection IWF. Thus any potential mechanical wear degradation would be detected by the implementation of IWE and IWF for those pressure retaining components and their supports.

* Evaluation is based on:
  (1) the 1992 Edition with the 1992 Addenda of Subsections IWE and IWL of Section XI of the ASME Code;
  (2) the 1989 Edition of Section XI of the ASME Code including Appendix VII, "Qualification of Nondestructive Examination Personnel for Ultrasonic Examination," and Appendix VIII (1989 Addenda), "Performance Demonstration for Ultrasonic Examination Systems;"
  (3) the final rule on 10 CFR 50.55a, Codes and Standards, dated August 8, 1996 (61 FR 41303); and
  (4) NUREG-1557, "Summary of Technical Information and Agreements from Nuclear Management and Resources Council Industry Reports Addressing License Renewal," dated October 1996.

Note (1) "Concrete interaction with aluminum" is not addressed in NUREG-1557 for BWR containment. However, this item is evaluated and considered not an issue for license renewal for PWR containment. Thus this item is considered not an issue for BWR containment.

APPENDIX A - IMPLEMENTATION HIGHLIGHTS OF SUBSECTIONS IWE AND IWL THROUGH 10 CFR 50.55a

Subsection IWE provides rules for inservice inspection, repair, and replacement of Class MC pressure retaining components and their integral attachments and of metallic shell and penetration liners of Class CC pressure retaining components and their integral attachments in light-water cooled power plants. Subsection IWL provides rules for inservice inspection and repair of the reinforced concrete and the post-tensioning systems of Class CC components. Licensees will be required to incorporate Subsection IWE and Subsection IWL into their inservice inspection (ISI) program. Licensees will be required to implement the containment examinations in accordance with Subsections IWE and IWL as endorsed by §50.55a by September 9, 2001.

In endorsing Subsections IWE and IWL, 10 CFR 50.55a sets forth additional requirements to assure that the critical areas of containments are periodically inspected to detect and take corrective action for defects that could compromise a containment's structural integrity. These additional requirements include:

(a) Four modifications specified in §50.55a(b)(2)(x) for examination of metal containments and the liners of concrete containments. These are: (1) Section 50.55a(b)(2)(x)(A) states that the licensee shall evaluate the acceptability of inaccessible areas of metal containments and the liners of concrete containments (Class MC), when conditions exist in accessible areas that could indicate the presence of or result in degradation to such inaccessible areas; (2) Section 50.55a(b)(2)(x)(B) permits alternative lighting and resolution requirements for remote visual examination of the containment; (3) Section 50.55a(b)(2)(x)(C) makes the examination of pressure retaining welds and pressure retaining dissimilar metal welds optional; and (4) Section 50.55a(b)(2)(x)(D) is added to provide an alternative sampling plan.

(b) Five modifications specified in §50.55a(b)(2)(ix) for examination of concrete containments. These modifications must be implemented when using Subsection IWL. Four of these issues are identified in Regulatory Guide 1.35, Revision 3, but are not addressed in the referenced Subsection IWL. The five modifications are: (1) Section 50.55a(b)(2)(ix)(A) requires that grease caps which are accessible be visually examined to detect grease leakage or grease cap deformation; (2) Section 50.55a(b)(2)(ix)(B) requires the preparation of an engineering evaluation report when consecutive surveillances indicate a trend of prestress loss to below the minimum prestress requirements; (3) Section 50.55a(b)(2)(ix)(C) requires that an evaluation be performed for instances of wire failure and slip of wires in anchorages; (4) Section 50.55a(b)(2)(ix)(D) addresses sampled sheathing filler grease and reportable conditions; and (5) Section 50.55a(b)(2)(ix)(E) requires that licensees evaluate the acceptability of inaccessible areas of concrete containments when conditions exist in accessible areas that could indicate the presence of or result in degradation to such inaccessible areas.

(c) One limitation specified in §50.55a(b)(2)(vi) for effective edition and addenda of Subsection IWE and Subsection IWL. It states that the 1992 Edition with the 1992 Addenda of Subsection IWE and Subsection IWL shall be used when performing containment examinations as modified and supplemented by the requirements described in §50.55a(b)(2)(ix) and §50.55a(b)(2)(x), respectively.

(d) One clarification specified in §50.55a(b)(2)(x)(E). It states that a general visual examination as required by Subsection IWE shall be performed once each period.

APPENDIX B - LIST OF PWR CONTAINMENT COMPONENTS

1. CONCRETE CONTAINMENTS (REINFORCED/PRESTRESSED)
- Concrete Dome
- Dome Reinforcing Steel
- Concrete Containment Wall Above Grade
- Containment Wall Reinforcing Steel Above Grade
- Concrete Containment Wall Below Grade
- Containment Wall Reinforcing Steel Below Grade
- Concrete Basemat
- Basemat Reinforcing Steel
- Containment Liner Interior Surface
- Containment Liner Above Grade Exterior Surface
- Containment Liner Below Grade Exterior Surface
- Basemat Liner Interior Surface
- Basemat Liner Exterior Surface
- Liner Anchors Above Grade
- Liner Anchors Below Grade

2. FREE-STANDING STEEL CONTAINMENT WITH FLAT BOTTOM & AN ICE CONDENSER
- Dome Shell Interior Surface
- Dome Shell Exterior Surface
- Cylindrical Shell Interior Surface
- Cylindrical Shell Exterior Surface
- Embedded Shell Region
- Concrete Basemat
- Basemat Reinforcing Steel
- Basemat Liner
- Liner Anchors

3. FREE-STANDING CYLINDRICAL & SPHERICAL STEEL CONTAINMENT WITH ELLIPTICAL BOTTOM
- Containment Shell Interior Surface
- Containment Shell Exterior Surface
- Embedded Shell Region
- Sand Pocket Region

4. CONCRETE CONTAINMENTS PRESTRESSED ONLY
- Prestressing Tendons

5. COMMON COMPONENTS
- Penetration Sleeves
- Penetration Bellows
- Personnel Airlock
- Equipment Hatches

Reference:
Page B-45 of NUREG-1557, "Summary of Technical Information and Agreements from Nuclear Management and Resources Council Industry Reports Addressing License Renewal," dated October 1996.

APPENDIX C - LIST OF BWR CONTAINMENT COMPONENTS

1. MARK I CONCRETE CONTAINMENTS
- Drywell Liner Interior Surface
- Drywell Liner Exterior Surface
- Torus Liner Interior Surface
- Torus Liner Interior Surface at Waterline
- Torus Liner Exterior Surface
- Liner Anchors
- Drywell Concrete
- Torus Concrete
- Drywell Concrete Reinforcing Steel
- Torus Concrete Reinforcing Steel
- Vent Lines
- Vent Line Bellows
- Vent Headers
- Downcomers and Bracing
- Vent System Supports
- Drywell Head

2. MARK I STEEL CONTAINMENTS
- Drywell Interior Surface
- Drywell Exterior Surface
- Drywell Head
- Embedded Shell Region
- Drywell Support Skirt
- Sand Pocket Region
- Torus Interior Surface
- Torus Interior Surface at Waterline
- Torus Exterior Surface
- Torus Ring Girder
- Vent Lines
- Vent Line Bellows
- Vent Header
- Downcomers and Bracing
- Drywell Exterior Shell with Compressible Material
- Vent System Supports
- Torus Seismic Restraints
- Torus Support Columns/Saddles
- ECCS Suction Header
- Ocean Plant with Uncoated CS Surfaces
- Uncoated Submerged CS Surfaces

3. MARK II CONCRETE CONTAINMENTS
- Drywell Liner Interior Surface
- Drywell Liner Exterior Surface
- Suppression Chamber Liner Interior Surface
- Suppression Chamber Liner Interior Surface at Waterline
- Suppression Chamber Liner Exterior Surface
- Liner Anchors
- Liner Region Shielded by Diaphragm Floor
- Containment Concrete

3. MARK II CONCRETE CONTAINMENTS (Continued)
- Concrete Containment Reinforcing Steel
- Drywell Head
- Downcomer Pipes and Bracing
- Concrete Basemat
- Basemat Liner
- Basemat Reinforcing Steel
- Prestressing Tendons and Ducts

4. MARK II STEEL CONTAINMENTS
- Drywell Interior Surface
- Drywell Exterior Surface
- Drywell Head
- Suppression Chamber Interior Surface
- Suppression Chamber Exterior Surface
- Suppression Chamber Interior Surface at Waterline
- Region Shielded by Diaphragm Floor
- Embedded Shell Region
- Sand Pocket Region
- Support Skirt
- Downcomer Pipes and Bracing
- Drywell Exterior Shell with Compressible Material
- Ocean Plant with Uncoated CS Surfaces
- Uncoated Submerged CS Surfaces

5. MARK III CONCRETE CONTAINMENTS
- Containment Liner Interior Surface
- Containment Liner Exterior Surface
- Suppression Chamber Liner or Cladding Interior Surface
- Suppression Chamber Liner Exterior Surface
- Concrete Containment Wall Above Grade
- Concrete Containment Wall Below Grade
- Concrete Dome
- Basemat Liner
- Concrete Basemat
- Liner Anchors
- Containment Wall Reinforcing Steel
- Dome Reinforcing Steel
- Basemat Reinforcing Steel

6. MARK III STEEL CONTAINMENTS
- Containment Shell Interior Surface
- Containment Shell Exterior Surface
- Suppression Chamber Shell Interior Surface
- Suppression Chamber Shell Exterior Surface
- Basemat Liner
- Liner Anchors
- Concrete Basemat
- Basemat Reinforcing Steel
- Concrete Fill in Annulus
- Embedded Shell Region

7. CONTAINMENT COMMON COMPONENTS
- Penetration Bellows
- Penetration Sleeves
- Dissimilar Metal Welds
- Personnel Airlock
- Equipment Hatches
- CRD Hatch

Reference:
Pages B-65, B-53, and B-52 of NUREG-1557, "Summary of Technical Information and Agreements from Nuclear Management and Resources Council Industry Reports Addressing License Renewal," dated October 1996.

NRC FORM 335
(2-89)
NRCM 1102,
3201, 3202

U.S. NUCLEAR REGULATORY COMMISSION

BIBLIOGRAPHIC DATA SHEET

(See instructions on the reverse)

1. REPORT NUMBER
(Assigned by NRC, Add Vol., Supp., Rev.,
and Addendum Numbers, If any.)

NUREG-1611

2 TITLE AND SUBTITLE

Aging Management of Nuclear Power Plant Containments
for License Renewal

3. DATE REPORT PUBLISHED

MONTH	YEAR
September	1997

4. FIN OR GRANT NUMBER

5 AUTHOR(S)

W. C. Liu, P. T. Kuo, S. S. Lee

6. TYPE OF REPORT

Regulatory

7. PERIOD COVERED *(Inclusive Dates)*

8. PERFORMING ORGANIZATION - NAME AND ADDRESS *(If NRC, provide Division, Office or Region, U S Nuclear Regulatory Commission, and mailing address; if contractor, provide name and mailing address.)*

Division of Reactor Program Management
Office of Nuclear Reactor Regulation
U.S. Nuclear Regulatory Commission
Washington, DC 20555-0001

9. SPONSORING ORGANIZATION - NAME AND ADDRESS *(If NRC, type "Same as above"; if contractor, provide NRC Division, Office or Region, U S Nuclear Regulatory Commission, and mailing address.)*

Same as 8. above.

10. SUPPLEMENTARY NOTES

11. ABSTRACT *(200 words or less)*

In 1990, the Nuclear Management and Resources Council (NUMARC), now the Nuclear Energy Institute (NEI), submitted for NRC review, the industry reports (IRs), NUMARC Report 90-01 and NUMARC Report 90-10, addressing aging management issues associated with PWR containments and BWR containments for license renewal, respectively. In 1996, the Commission amended 10 CFR 50.55a to promulgate requirements for inservice inspection of containment structures. This rule amendment incorporates by reference the 1992 Edition with the 1992 Addenda of Subsections IWE and IWL of the ASME Code addressing the inservice inspection of metal containments/liners and concrete containments, respectively. The purpose of this report is to reconcile the technical information and agreements resulting from the NUMARC IR reviews which are generally described in NUREG-1557 and the inservice inspection requirements of subsections IWE and IWL as promulgated in §50.55a for license renewal consideration. This report concludes that Subsections IWE and IWL as endorsed in §50.55a are generally consistent with the technical agreements reached during the IR reviews. Specific exceptions are identified and additional evaluations and augmented inspections for renewal are recommended.

12. KEY WORDS/DESCRIPTORS *(List words or phrases that will assist researchers in locating the report.)*

aging management
license renewal
containment structures
Subsections IWE and IWL
10 CFR 50.55a
10 CFR Part 54

13 AVAILABILITY STATEMENT

unlimited

14 SECURITY CLASSIFICATION

(This Page)

unclassified

(This Report)

unclassified

15. NUMBER OF PAGES

16. PRICE

www.ingramcontent.com/pod-product-compliance
Lightning Source LLC
Chambersburg PA
CBHW081606170526
45166CB00009B/2854